Saturn

Uranus

Neptune

Pluto

Neptune and Pluto

Editor-in-chief: Paul A. Kobasa

Writer: Robert N. Knight

Editors: Jeff De La Rosa, Lisa Kwon, Maureen Liebenson,
Christine Sullivan

Research: Mike Barr, Jacqueline Jasek, Barbara Lightner,
Andy Roberts, Loranne Shields

Graphics and Design: Sandra Dyrlund, Charlene Epple,
Brenda Tropinski

Photos: Kathy Creech, Tom Evans

Permissions: Janet Peterson

Indexing: David Pofelski

Proofreading: Anne Dillon

Text processing: Curley Hunter, Gwendolyn Johnson

Pre-press and Manufacturing: Carma Fazio, Anne Fritzinger,
Steven Hueppchen, Madelyn Underwood

World Book, Inc.
233 N. Michigan Avenue
Chicago, IL 60601
U.S.A.

Library of Congress Cataloging-in-Publication Data

Neptune and Pluto.
    p. cm. -- (World Book's solar system &
space exploration library)
    Includes bibliographical references and index.
    ISBN 0-7166-9507-3
    1. Neptune (Planet)--Juvenile literature. 2. Pluto
(Planet)--Juvenile literature.
I. World Book, Inc. II. Series.
QB691.N44 2006
523.48--dc22
                                        2005030449

ISBN (set) 0-7166-9500-6

Printed in the United States of America

1 2 3 4 5 6 7 8 09 08 07 06

For information about other World Book publications,
visit our Web site at http://www.worldbook.com or call
1-800-WORLDBK (967-5325).

For information about sales to schools and libraries,
call 1-800-975-3250 (United States);
1-800-837-5365 (Canada).

Picture Acknowledgments: Back Cover: NASA; © Mark Garlick, SPL/Photo Researchers; NASA; Johns
Hopkins University Applied Physics Laboratory/Southwest Research Institute; Inside Front Cover: © John
Gleason, Celestial Images.

© Vanni/Art Resource 29; © Calvin J. Hamilton 13; Johns Hopkins University Applied Physics
Laboratory/Southwest Research Institute 59; Lunar and Planetary Institute 41; © Mary Evans Picture Library
57; NASA 11, 15, 25, 33, 49; NASA/JPL 23; NASA, L. Sromovsky, and P. Fry, University of Wisconsin-
Madison 17; © SPL/Photo Researchers 51; © Chris Butler, Photo Researchers 39, 55, 61; © John R. Foster,
Photo Researchers 43; © Mark Garlick, SPL/Photo Researchers 53; © David A. Hardy, /SPL/Photo
Researchers 27.

Illustrations: Inside Back Cover: WORLD BOOK illustration by Steve Karp; Front Cover & 1, 3, 9, 21, 37,
47: WORLD BOOK illustrations by Paul Perreault; WORLD BOOK illustrations by Precision Graphics 7, 35;
WORLD BOOK illustration 31.

Astronomers use different kinds of photos to learn about objects in space—such as planets. Many photos
show an object's natural color. Other photos use false colors. Some false-color images show types of light
the human eye cannot normally see. Others have colors that were changed to highlight important features.
When appropriate, the captions in this book state whether a photo uses natural or false color.

WORLD BOOK'S

SOLAR SYSTEM & SPACE
EXPLORATION LIBRARY

Neptune and Pluto

World Book, Inc.
a Scott Fetzer company
Chicago

Contents

NEPTUNE

> If a word is printed in **bold letters that look like this,** that word's meaning is given in the glossary on page 63.

PLUTO

Where Is Neptune?

Neptune is the eighth **planet** from the sun. Neptune's **orbit** lies between the orbits of Uranus, its inner neighbor, and Pluto, its outer neighbor. Once about every 250 years, however, Pluto's **elliptical** orbit dips inside Neptune's orbit for a span of about 20 years. Then, Neptune is the farthest planet from the sun. This change in position between Neptune and Pluto last happened from January 1979 to February 1999, so Pluto is again the farthest planet from the sun and will be for a long time.

Neptune's average distance from the sun is about 2.8 billion miles (4.5 billion kilometers). That's about 30 times farther from the sun than Earth is.

Neptune is one of the planets in our **solar system** that **astronomers** call the outer planets. The other outer planets are Jupiter, Saturn, Uranus (*YUR uh nuhs* or *yu RAY nuhs*), and Pluto.

Planet Locator

Note: The size of the sun and planets and the distance between planets in this diagram are not to scale.

Pluto

Neptune

Uranus

Saturn

Jupiter

Mars

Earth

Venus

Mercury

Sun

Neptune's symbol (top left) and a diagram showing the planet's location in the solar system

How Big Is Neptune?

Neptune is the fourth-largest **planet** in our **solar system.** Only Jupiter, Saturn, and Uranus are larger. However, Uranus is only slightly larger than Neptune. If the planets were put side by side, you would have to look closely to see there is a difference in size.

The **diameter** of Neptune at its **equator** is 30,775 miles (49,528 kilometers). That is nearly four times the diameter of Earth. Next to Earth, Neptune would look like a giant planet. If Neptune were hollow, it would take about 58 planets the size of Earth to fill it up. But it would take nearly 10,000 Plutos to fill up Neptune!

An artist's drawing comparing the size of Neptune and Earth

Neptune's diameter
30,775 miles
(49,528 kilometers)

Earth's diameter
7,926 miles
(12,756 kilometers)

What Does Neptune Look Like?

Neptune is far from Earth—so far that we cannot see it in the sky without a telescope. But when seen through a telescope, Neptune appears as a bright point in the sky, like a star.

When photographed from space by a **probe,** Neptune looks like a giant blue ball with tiny wisps of white clouds. The planet has a belt of six rings around its middle, but the rings are so faint that they do not show up in most photographs. Neptune's bright blue color is caused by **methane** gas at the top of its **atmosphere.**

Neptune's high, thin, streaky white clouds look like clouds on Earth that are called cirrus clouds. Cirrus clouds are made of ice crystals that form very high in Earth's atmosphere.

Neptune in a
natural-color photo

What Makes Up Neptune?

Scientists label the four largest **planets** in the **solar system** the **gas giants.** These planets are Jupiter, Saturn, Uranus, and Neptune. All these planets are made up of large amounts of gas and liquid.

Scientists believe that Neptune is a huge ball of highly pressurized gas and liquid. They think that Neptune is made up mostly of **hydrogen, helium,** water, and **silicates.** Unlike Earth, Neptune has no solid ground.

The part of Neptune that we can see consists of several cloud layers. Deeper in, there is a layer of compressed gases—mostly hydrogen and helium gases. Farther inside the planet, in the layer known as the **mantle,** the gases blend into a layer of liquid. Some scientists think that this layer of liquid might be superheated water that would boil away if it could, but the **pressure** of the surrounding gases keeps it from doing so. Inside the mantle is most likely a solid **core** made of ice and rock.

The Interior of Neptune

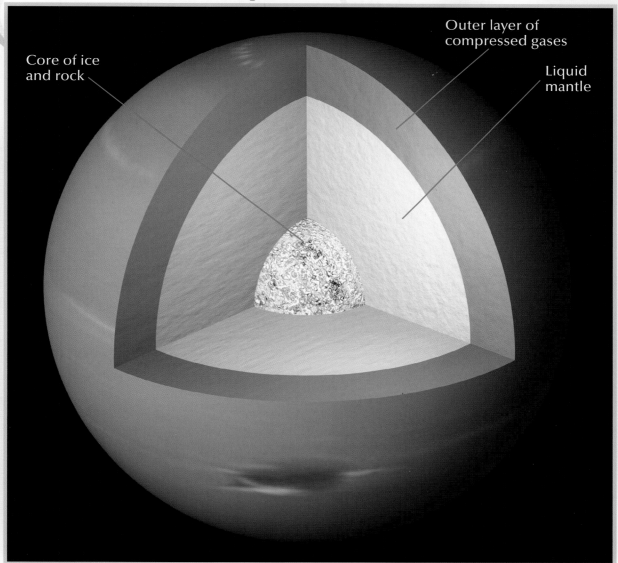

Core of ice
and rock

Outer layer of
compressed gases

Liquid
mantle

What Is Neptune's Atmosphere Like?

Neptune is surrounded by thick layers of clouds. The **atmosphere** of Neptune seems to be layered. Clouds made up mainly of **methane** surround Neptune and give the planet its blue color. The clouds farthest from Neptune's surface consist mainly of frozen methane.

Darker cloud layers deeper inside Neptune seem to be made of **hydrogen sulfide.** This colorless gas has the odor of rotten eggs and, at high levels, it can be dangerous or even deadly to humans. Neptune's atmosphere also contains **hydrogen** and **helium.**

Clouds on Neptune in a natural-color photo

What Is the Weather on Neptune?

The weather on Neptune is cloudy, windy, and cold! The average temperature of Neptune's outer cloud layer is −355 °F (−215 °C).

Neptune is a world of violent winds. These fierce winds blow at speeds as high as 900 miles (1,450 kilometers) per hour! Earth's strongest winds blow only about one-third that speed. The winds of Neptune blow the thick cloud layers around that planet very fast.

Scientists have found evidence of changing seasons on Neptune. The tilt of Neptune's **axis** is similar to the tilt of Earth's axis, so Neptune has four seasons, as does Earth. When scientists recently studied photographs of the planet taken over a span of several years, they found that clouds in the southern half of Neptune have become much brighter—evidence that it is now spring on that part of the planet. On Neptune, one season is not, as on Earth, three months long—instead, it lasts for more than 40 years! Spring may last for around another 20 years on Neptune.

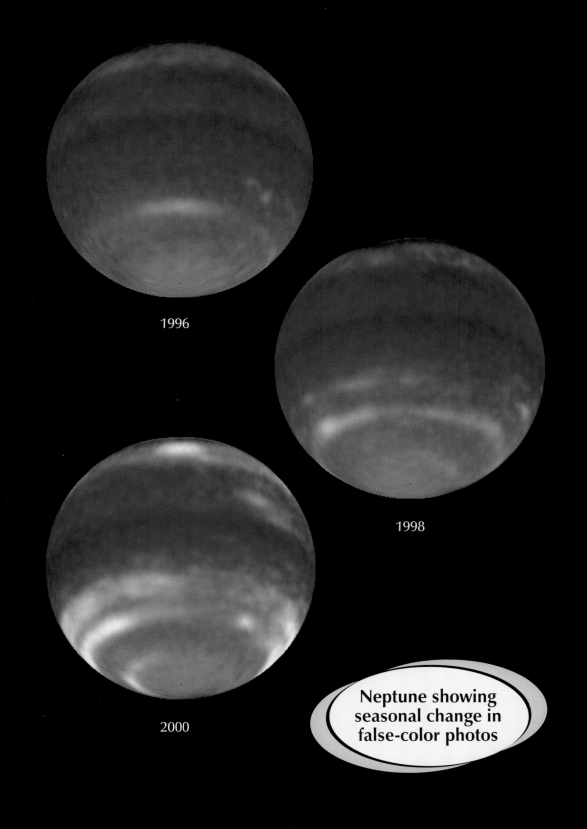

1996

1998

2000

Neptune showing
seasonal change in
false-color photos

How Does Neptune Compare with Earth?

Although Neptune has about 17 times more **mass** than Earth, its average **density** is only about 1.5 times that of Earth's. Neptune's surface **gravity** is slightly greater than Earth's. If you weighed 100 pounds on Earth, you would weigh about 112 pounds on Neptune.

Earth is close enough to the light-giving sun to absorb much of that heat and keep out the frigid conditions of space. But Neptune is on the outer fringes of our **solar system.** It receives a tiny fraction of the sunlight that bathes Earth. If you could view the sun from Neptune, it would look 900 times dimmer than here on our Earth.

How Do They Compare?

	Earth ⊕	Neptune ♆
Size in diameter (at equator)	7,926 miles (12,756 kilometers)	30,775 miles (49,528 kilometers)
Average distance from sun	About 93 million miles (150 million kilometers)	About 2.8 billion miles (4.5 billion kilometers)
Length of year (in Earth days)	365.256	60,189
Length of day (in Earth time)	24 hours	16 hours 7 minutes
What an object would weigh…	If it weighed 100 pounds on Earth…	…it would weigh about 112 pounds on Neptune.
Number of moons	1	at least 13
Rings?	No	Yes
Atmosphere	Nitrogen, oxygen, argon	Hydrogen, helium, methane, acetylene

What Are the Orbit and Rotation of Neptune Like?

Like the other **planets** in our **solar system,** Neptune travels in an **orbit** around the sun that is **elliptical.** However, Neptune's orbit is more nearly round than the orbits of all the other planets except Venus.

On average, Neptune is about 2.8 billion miles (4.5 billion kilometers) from the sun. Because the planet is so very far from the sun, it takes Neptune about 165 Earth years to complete one orbit. In fact, in the time since Neptune was first discovered by **astronomers** in the mid-1800's, the planet has not yet completed one orbit. It will do so in 2011. With a **year** that is around 165 Earth years long, New Year's celebrations would be few and far between on Neptune!

Neptune rotates (spins around) on its **axis,** making one complete turn, or rotation, every 16 hours and 7 minutes in Earth time. That's the length of a **day** on Neptune.

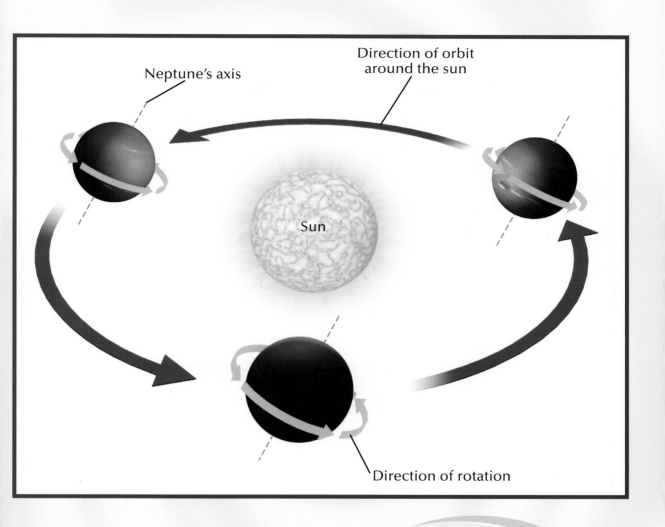

Neptune's axis

Direction of orbit around the sun

Sun

Direction of rotation

A diagram showing the orbit and rotation of Neptune

What Makes Up Neptune´s Rings?

Like the other **gas giants** in our **solar system,** Neptune has a system of rings. Scientists have counted six rings circling the planet's **equator.**

Neptune's rings are probably made of dust and are very faint, but three arcs, or sections, of the outer ring shine more brightly than other parts of the ring. Scientists believe that dust may be concentrated more thickly in these bright sections. One of Neptune's **moons,** Galatea *(gal uh TEE uh),* **orbits** the **planet** just inside the outer ring. The pull of its **gravity** probably concentrates dust in the three bright arcs.

We did not know anything about rings on Neptune until the 1980's, when **astronomers** watched the planet pass in front of a star. The starlight "winked" when it crossed a ring. The Voyager 2 **probe**—launched by the National Aeronautics and Space Administration (NASA) in 1977—confirmed the existence of the rings in 1989.

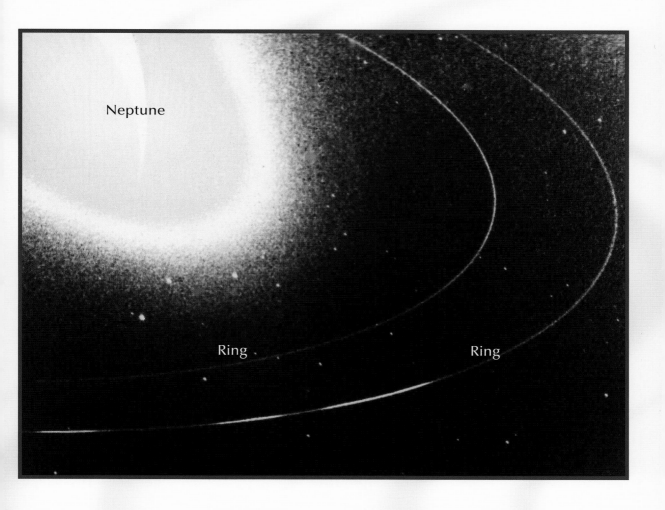

Neptune

Ring

Ring

23

How Many Moons Does Neptune Have?

Neptune has at least 13 **moons.** The three largest moons are Triton *(TRY tuhn),* Proteus *(PROH tee uhs),* and Nereid *(NIHR ee ihd).* Triton is one of the larger moons in our **solar system,** but Neptune's other moons are quite small.

Proteus, Neptune's second-largest moon, is dark and quite close to the planet. **Astronomers** using telescopes on Earth overlooked this small moon, but images sent by Voyager 2 when it visited Neptune in 1989 allowed scientists to discover Proteus. Nereid, Neptune's third-largest moon, has an **orbit** that is the most **elliptical** of any moon in the solar system. Proteus, Nereid, and Neptune's other small moons are all less than 300 miles (480 kilometers) in **diameter.**

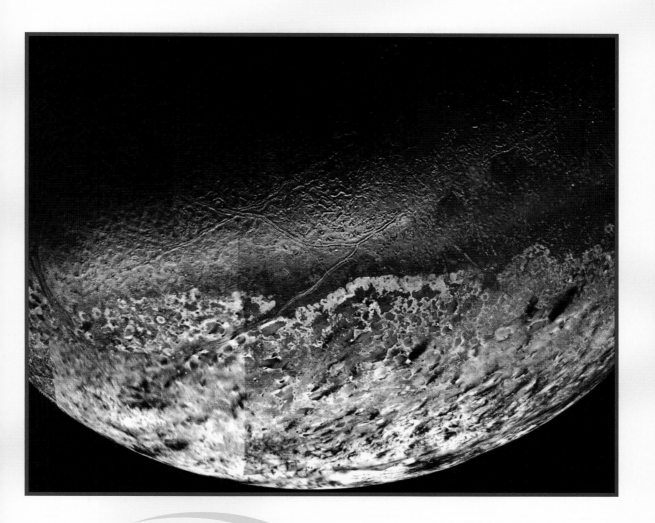

Neptune's moon Triton in
a natural-color photo

What Is Unusual About Neptune's Moon Triton?

Triton, which is Neptune's largest **moon,** has a **diameter** of about 1,680 miles (2,700 kilometers). It is one of the larger moons in our **solar system.** Only Jupiter's four largest moons, Earth's moon, and Saturn's moon Titan are larger.

Like Saturn's Titan—and unlike most other moons—Triton has an **atmosphere.** It is made mainly of **nitrogen** gas. Triton has **geysers**—streams of nitrogen and other material that spurt occasionally from below its surface. Triton is the coldest body in the solar system, with a surface temperature of about −390 °F (−235 °C).

Triton **orbits** Neptune in the direction opposite to Neptune's rotation. That's why scientists think Triton was captured by Neptune's **gravity** long after Neptune formed. Triton's orbit is gradually dropping closer to Neptune. Millions of years from now, it may break up and form a new ring around Neptune.

An artist's drawing imagining the surface of Triton, with Neptune in the background

How Did Neptune Get Its Name?

People in ancient times knew nothing about the **planet** Neptune. In the 1800's, several **astronomers** noticed an unusual movement in the **orbit** of the planet Uranus and predicted the existence of another planet orbiting beyond Uranus. German astronomer Johann G. Galle *(GAHL uh)* and his assistant, Heinrich L. d'Arrest *(duh REST),* first viewed the planet we now know as Neptune through a telescope in 1846. However, the scientists who had correctly predicted Neptune's location—John C. Adams of England and Urbain J. J. Leverrier *(luh VEHR ee ay)* of France—are given credit for the discovery.

The scientists named the new planet for Neptune, the Roman god of the sea. The name fits well, for the planet looks blue, like a lake or sea. Neptune's **moons** are also named for mythical sea beings. Triton, for example, was the name the ancient Greeks used for a sea god in the form of a merman—a humanlike creature with the tail and lower body of a fish.

The symbol for the planet Neptune (see page 7) is a three-pronged spear that the god is said to have carried.

What Space Missions Have Visited Neptune?

A **probe** from NASA's Voyager missions, Voyager 2, was launched from Earth in 1977. On August 25, 1989, the Voyager 2 probe passed within about 3,000 miles (4,850 kilometers) of the cloud-covered surface of Neptune. That was 3 ½ years after the probe's **fly-by** of Uranus, exactly 8 years after the probe's visit to Saturn, and 10 years after its Jupiter encounter.

Voyager 2 sent a large amount of information about Neptune back to Earth, which greatly expanded our knowledge about that planet. The probe found six new **moons** around Neptune. Previously, only two—Triton and Nereid—had been known. Voyager 2 also increased **astronomers'** limited knowledge of Neptune's faint rings.

Voyager 2 data also revealed much about Triton, Neptune's largest and most interesting moon. The probe returned photographs showing **geysers** spurting **nitrogen** gas and dust into Triton's thin **atmosphere.**

An artist's drawing of
Voyager 2

Could There Be Life on Neptune or Its Moons?

Scientists believe there is very little chance of finding any kind of life on Neptune or its **moons.** Neptune's **atmosphere** is made up of gases that would be quite poisonous if breathed by most of the living things we know about. And, the temperature at the cloudtops is an incredibly frigid −355 °F (−215 °C)!

Deep inside Neptune, there may be an ocean of water, but if so, it is probably boiling hot. And the **pressure** inside Neptune may be great enough to crush spaceships, much less living things.

Neptune's only large moon with any atmosphere is Triton. But it is very hard to imagine any kind of life that could exist at such extremely cold temperatures. The most frigid places on Earth at Antarctica would seem warm and inviting compared to Triton.

There may be life somewhere in the universe other than on Earth. But it is not likely to be found on or near Neptune.

Neptune (top) and Triton
in a false-color photo

Where Is Pluto?

Pluto is the **planet** in our **solar system** that is farthest from the sun. Pluto's average distance from the sun is over 3.6 billion miles (5.9 billion kilometers). That makes it about 39 times farther from the sun than Earth is. Pluto travels in a lopsided **orbit** around the sun, however, so it comes inside the orbit of Neptune for relatively short periods of time. During those times, Neptune is farther from the sun than Pluto is. It will be more than 200 years before Pluto and Neptune cross orbits again. Until then, Pluto will be the planet farthest from the sun.

Pluto is the last of what **astronomers** call the outer planets. The other outer planets are Jupiter, Saturn, Uranus, and Neptune.

Planet Locator

Note: The size of the sun and planets and the distance between planets in this diagram are not to scale.

Pluto

Neptune

Uranus

Saturn

Jupiter

Mars

Earth

Venus

Mercury

Sun

Pluto's symbol (top left) and a diagram showing the planet's location in the solar system

How Big Is Pluto?

Pluto is by far the smallest **planet** in our **solar system**. **Astronomers** estimate its **diameter** at the **equator** to be 1,485 miles (2,390 kilometers). The next-smallest planet is Mercury, which has a diameter of 3,032 miles (4,879 kilometers). In fact, even Earth's **moon** is larger than Pluto.

If you could put Earth and Pluto side by side, Earth would dwarf Pluto. The diameter of Earth is 7,926 miles (12,756 kilometers). That's about 5 ½ times Pluto's diameter. But, if Earth were hollow, it would take about 175 Plutos to fill it.

An artist's drawing
comparing the size of Pluto
and Earth

Pluto's diameter
1,485 miles
(2,390 kilometers)

Earth's diameter
7,926 miles
(12,756 kilometers)

What Does Pluto Look Like?

Pluto is a small **planet.** The closest it ever gets to Earth is about 2.7 billion miles (4.3 billion kilometers) away. That is so far away that in even the most powerful telescopes Pluto looks like a blurry disk from Earth.

From what **astronomers** have been able to gather from such tools as the Hubble Space Telescope, Pluto appears to have a brownish color. It has bright splotches that are probably **polar caps.** Astronomers can also make out dark spots scattered across Pluto's surface. For a clearer image of Pluto, however, we will have to wait for a space **probe** to visit the distant planet.

An artist's drawing
of Pluto

What Makes Up Pluto?

Unlike Neptune and the other outer **planets,** Pluto is a rocky ball with solid ground—like Earth and the other inner planets, Mercury, Venus, and Mars. Because Pluto is so far away and we have so little scientific data about it, scientists can only make educated guesses about its composition. But they do have some clues.

Scientists know that Pluto has less **density** than Earth. That means that the matter that makes up Pluto is packed much less tightly than Earth's matter is.

Scientists think much of Pluto could be made of water ice. The planet may have a **core** of solid metals and rock surrounded by a thick layer of ice. If so, that would explain why Earth has more density than Pluto does. Rock is more dense than water, which is why a rock sinks in water. Ice is less dense than water, and that is why icebergs float on water. So, if Earth were much rockier than Pluto was, that would explain the difference in density between the two planets.

The Interior of Pluto

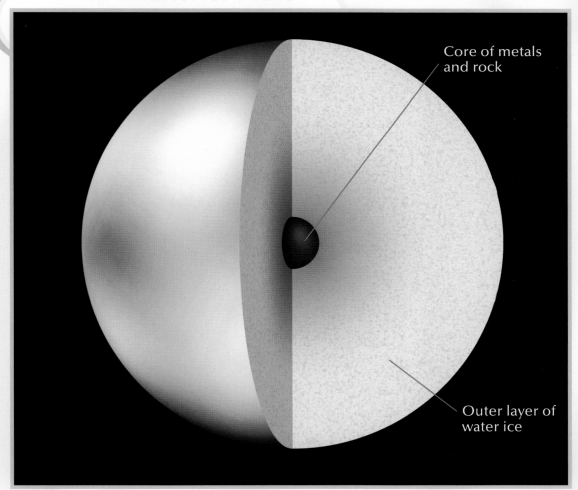

Core of metals and rock

Outer layer of water ice

Does Pluto Have an Atmosphere and Weather?

Pluto has an **atmosphere,** but it is very, very thin. Scientists think that Pluto's atmosphere is made up mostly of **methane** gas. We know nothing about wind currents or storms on Pluto. However, Pluto does have changes in its atmosphere that we would call weather.

When Pluto is closest to the sun in its **orbit,** some of the frozen materials on its surface turn to gas and this adds material to the atmosphere. On the other hand, when Pluto is farthest from the sun, materials in the atmosphere freeze and fall back to the surface as ice.

Scientists also think that some of Pluto's atmosphere "bleeds" into space when the planet is nearest the sun. Why doesn't Pluto's atmosphere eventually all bleed away? Scientists don't know the answer to this question. Perhaps Pluto has **geysers** or volcanoes that shoot material into the atmosphere from inside the planet.

An artist's drawing of
the surface of Pluto with
Charon rising and a
distant sun

How Does Pluto Compare with Earth?

In one important way, Pluto is more like Earth than its distant neighbor Neptune is. Unlike Neptune—but like Earth—Pluto is a solid ball. It has solid ground and an **atmosphere** that starts at that ground level.

However, in many other ways, Pluto is quite different from Earth. Pluto is an icy, frigid world. Temperatures there hover around −375 °F (−225 °C). Pluto's thin atmosphere would seem like empty space compared to Earth's thick blanket of air. Earth's atmosphere is many thousands of times more **dense** than Pluto's atmosphere. Pluto is far too distant from the sun to be very much like Earth.

Because Pluto has much less **mass** than Earth does, Pluto's surface **gravity** is much less than Earth's. If you weighed 100 pounds on Earth, you would only weigh about 6.7 pounds on Pluto.

How Do They Compare?

	Earth ⊕	Pluto ♇
Size in diameter (at equator)	7,926 miles (12,756 kilometers)	1,485 miles (2,390 kilometers)
Average distance from sun	About 93 million miles (150 million kilometers)	About 3.6 billion miles (5.9 billion kilometers)
Length of year (in Earth days)	365.256	90,465
Length of day (in Earth time)	24 hours	153 hours 17 minutes
What an object would weigh…	If it weighed 100 pounds on Earth…	…it would weigh about 6.7 pounds on Pluto.
Number of moons	1	At least 1
Rings?	No	No
Atmosphere	Nitrogen, oxygen, argon	Methane, nitrogen

What Are the Orbit and Rotation of Pluto Like?

Pluto's **orbit** around the sun is the most **elliptical** of all the **planets.** At its closest point to the sun, Pluto is about 2.7 billion miles (4.3 billion kilometers) away. At its farthest distance, the planet is about 4.7 billion miles (7.5 billion kilometers) from the sun.

Pluto is so far from the sun that it takes about 248 Earth years to go around the sun once. That's one **year** on Pluto. For about 20 Earth years out of that 248-year orbit time, Pluto dips inside the orbit of Neptune, its nearest neighbor. That happened last between 1979 and 1999. It will not happen again until the 2200's.

Pluto rotates (spins around) on its **axis** about one time every six Earth **days.** So, a day on Pluto lasts for a very long time!

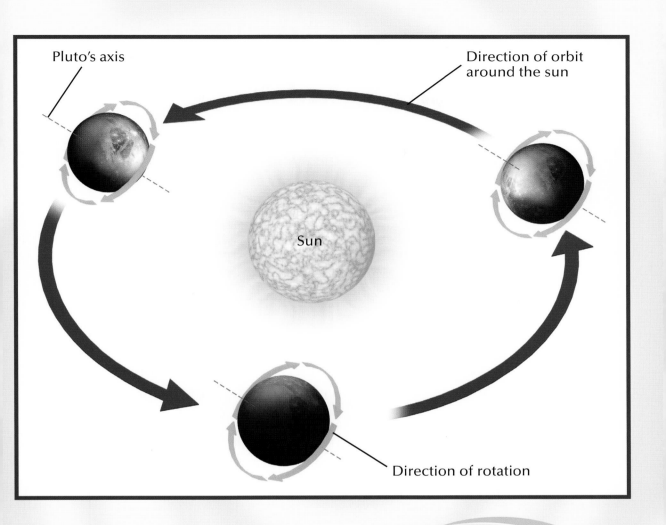

Pluto's axis

Direction of orbit
around the sun

Sun

Direction of rotation

A diagram showing
the orbit and rotation
of Pluto

What Is the Surface of Pluto Like?

Is the surface of Pluto rough and rocky, or is it smooth and flat? The answer to these questions will have to await stronger telescopes than we have at present or a space mission to Pluto. For now, we just do not know very much about Pluto.

There is a little we do know, however. Images of Pluto seen through powerful telescopes reveal many bright and dark areas. Some of the most detailed of these images have come from the Hubble Space Telescope. Scientists think the bright areas, including **polar caps,** are ice or frost. The caps may contain frozen **methane, nitrogen,** or **carbon monoxide.** These materials are normally gases on Earth, but on frigid Pluto they are frozen much of the time.

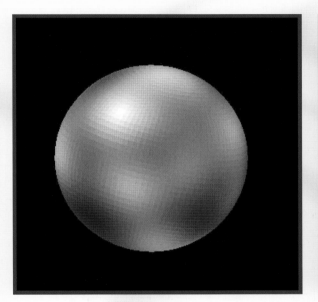

Pluto in two black-and-white photomosaics

How Many Moons Does Pluto Have?

Pluto has at least one **moon.** That moon is called Charon *(KAIR uhn)*. Charon is about half the size of Pluto. In our **solar system,** it is quite unusual for a **planet** to have a moon so large. Most moons are much, much smaller than their parent planet. Also, Pluto and Charon are very close to each other in distance—only about 12,200 miles (19,600 kilometers) apart. Our own moon is about 239,000 miles (384,000 kilometers) from Earth.

Scientists think that Charon's surface is covered by dirty water ice. Charon's surface is much less reflective, or shiny, than Pluto's surface, and it does not have contrasting bright and dark spots like Pluto. Charon probably does not have an **atmosphere,** either.

It may be that Pluto has other moons in addition to Charon. In October 2005, NASA announced that the Hubble Space Telescope had sent images of two objects which **astronomers** believed to be moons of Pluto. These objects, designated S/2005 P1 and S/2005 P2, are two to three times as far from Pluto as Charon is and are much smaller than Charon.

An artist's drawing of
Pluto (right) and Charon
(above left)

Are Pluto and Charon a Double Planet?

Some scientists think that Pluto and Charon are actually a double **planet,** rather than a planet with a **moon.** You have read that Charon is unusually large for a moon—when compared to Pluto's size. Because of the way Pluto and Charon move in their **orbits** through space, it is hard to say which body is orbiting which. All of these factors make the relationship between Pluto and Charon very unusual.

Both of these bodies keep the same face forever toward each other as they rotate. If you were on Pluto, you would always see Charon in the same place in the sky. And if you were on Charon, you would always see Pluto in the same part of the sky. It's as if Pluto and Charon are locked in a dance together across the heavens! In our **solar system,** no other planet and moon behave like the odd couple, Pluto and Charon.

An artist's drawing of Pluto (bottom) and Charon

Is Pluto Even a Planet?

Scientists are studying a wide region at the edge of our **solar system** called the **Kuiper belt,** which is named for the Dutch-born American **astronomer** Gerard P. Kuiper *(KY pur)*. This region is a band of many small-to-medium-sized objects called Kuiper belt objects—or KBO's. Many scientists believe that most KBO's are rocky and icy—like Pluto and also like Charon.

Scientists have long wondered why tiny, rocky Pluto occupies a place in the outer solar system along with the **gas giants**—Jupiter, Saturn, Uranus, and Neptune. Now some scientists think that Pluto and Charon are KBO's and belong to the Kuiper belt—instead of the sun's family of **planets.**

In 2005, astronomers announced the discovery of an object orbiting the sun—far beyond the **orbit** of Pluto—that some called the 10th planet. This Kuiper belt object is probably larger than Pluto. Until it can be officially named, the object's number designation is 2003 UB313. This object created even more confusion about whether Pluto should keep its status as a planet.

An artist's drawing
of Kuiper belt objects

How Did Pluto Get Its Name?

American **astronomer** Clyde W. Tombaugh discovered Pluto in 1930 while using the telescope of the Lowell Observatory in Flagstaff, Arizona. After observatory officials announced the discovery, people from all over the world sent in suggestions for the name of the **planet.** Venetia Burney, an 11-year-old girl in England, suggested the name "Pluto." In ancient Roman mythology, Pluto was the god of the underworld. The astronomy officials adopted Venetia Burney's idea.

American astronomer James W. Christy discovered Pluto's moon Charon in 1978. Christy chose the name "Charon," which—like Pluto—comes from Roman mythology. Charon was the boatman who ferried the spirits of the dead across the river to the underworld— where Pluto ruled.

An engraving of the Roman god Pluto (upper right, seated with his queen) and the god Charon (left)

Is There a Space Mission Planned to Visit Pluto?

There is still a lot we do not know about Pluto. Our best telescopes on Earth can produce only small, blurry images of the planet. That's one reason why in the early 2000's, NASA began planning the New Horizons space mission to Pluto, Charon, and the **Kuiper belt** beyond.

The New Horizons spacecraft weighs about 1,000 pounds (about 450 kilograms) and is about the size of a grand piano. It is loaded with scientific instruments that can take pictures and temperatures, analyze chemicals, and make surface maps, among many other things.

As of mid-2005, the New Horizons spacecraft was scheduled to blast off in early 2006. If it stays on schedule, the **probe** will reach Pluto and Charon in 2015. Students who are in first grade when the spacecraft is expected to blast off will be nearly finished with high school when New Horizons starts sending pictures back from Pluto!

An artist's drawing of the New Horizons spacecraft

Could There Be Life on Pluto or Charon?

Pluto is a solid ball with a thin blanket of air. The planet has solid ground to stand on, unlike Neptune, whose surface consists of gas and liquid. So, could there be life on Pluto? Scientists say the odds seem very much against it.

Pluto is so far from the sun that it resembles a dark deep-freeze. The sunlight is probably too feeble to support plantlike life forms, for example. Pluto's extremely thin **atmosphere** presents another stumbling block for life. Life forms we know of need some sort of atmosphere in order to live.

Charon seems even less inviting as a place for living things. It is at least as cold as Pluto, and it does not even have an atmosphere.

An artist's drawing of the surface of Pluto, with a dim sun (at left) and Charon (right)

FUN FACTS About NEPTUNE & PLUTO

★ Triton, Neptune's largest **moon,** was discovered only three weeks after the planet itself, in 1846.

★ Gerard P. Kuiper, the **astronomer** for whom the **Kuiper belt** is named, discovered Neptune's moon Nereid in 1949. It was the first moon of Neptune to be discovered in 103 years—but it would not be the last.

★ Sunlight on Neptune is 900 times dimmer than on Earth, yet Neptune has violent weather. Scientists are not really sure why—they would predict that a **planet** that receives so little of the sun's energy should have much less active weather.

★ Neptune's moon Triton is colder than the planet Pluto—even though Pluto is farther from the sun most of the time. Triton has an icy, shiny surface that reflects most of the sunlight that reaches it back into space, which keeps Triton very, very chilly!

★ Pluto is so small, if you placed it over the United States, the planet would be only about half the width of the U.S.

Glossary

astronomer A scientist who studies stars and planets.

atmosphere The mass of gases that surrounds a planet.

axis In planets, the imaginary line about which the planet seems to turn, or rotate. (The axis of Earth is an imaginary line through the North Pole and the South Pole.)

carbon monoxide A compound formed of carbon and oxygen.

core The center part of the inside of a planet.

day The time it takes a planet to rotate (spin) once around its axis and come back to the same position in relation to the sun.

density The amount of matter in a given space.

diameter The distance of a straight line through the middle of a circle or a thing shaped like a ball.

elliptical Having the shape of an ellipse, which is like an oval or flattened circle.

equator An imaginary circle around the middle of a planet.

fly-by The flight of a space vehicle close to a planet or body in space.

gas giant Any of four planets—Jupiter, Saturn, Uranus, and Neptune—made up mostly of gas and liquid.

geyser A stream that spurts hot water or gases and other material with explosive force from time to time.

gravity The effect of a force of attraction that acts between all objects because of their mass (that is, the amount of matter the objects have).

helium The second most abundant chemical element in the universe.

hydrogen The most abundant chemical element in the universe.

hydrogen sulfide A compound formed of hydrogen and sulfur.

Kuiper belt A ring of small objects orbiting in the outer solar system, beyond Neptune. Scientists believe that many comets are objects from the Kuiper belt.

mantle The area of a planet between the crust and the core.

mass The amount of matter a thing contains.

methane A compound formed of the chemical elements carbon and hydrogen.

moon A smaller body that orbits a planet.

nitrogen A nonmetallic chemical element.

orbit The path that a smaller body takes around a larger body, for instance, the path that a planet takes around the sun. Also, to travel in an orbit.

planet A large, round body in space that orbits a star and shines with light reflected from that star.

polar cap A white or bright area on the poles of some planets, including Mars and Pluto, that resembles ice or snow.

pressure The force caused by the weight of a planet's atmosphere as it presses down on the layers below it.

probe An unpiloted device sent to explore space. Most probes send data (information) from space.

silicate One of a group of minerals that contain silicon, oxygen, and one or more metallic elements.

solar system A group of bodies in space made up of a star and the planets and other objects orbiting around that star.

year The time it takes a planet to complete one orbit around the sun.

Index

For more information about Neptune and Pluto, try these resources:

The Far Planets by Robin Kerrod, Raintree, 2002
Neptune:
Neptune, by Carmen Bredeson, Franklin Watts, 2003
Neptune, by Seymour Simon, Sagebrush, 1999
Pluto:
A Double Planet?: Pluto and Charon (Isaac Asimov's New Library of the Universe), by Isaac Asimov, Gareth Stevens Press, 1996
Postcards from Pluto: A Tour of the Solar System, by Loreen Leedy, Holiday House, 1996

Neptune:
http://nssdc.gsfc.nasa.gov/planetary/planets/neptunepage.html
http://www.windows.ucar.edu/tour/link=/neptune/neptune.html
Pluto:
http://nssdc.gsfc.nasa.gov/planetary/planets/plutopage.html
http://pluto.jhuapl.edu/